# OCEAN

by Sheila Rivera

nneapolis

What is an **ocean?**

It is a large body of water.

An ocean is bigger than a lake.

There are five oceans
on Earth.

An ocean is a kind of **habitat.**

A habitat is where plants
and animals live.

Plants that live in the ocean are called **seaweed.**

There are many kinds of seaweed.

Some of the animals that live in the ocean are **fish.**

A shark is a fish.

Eels are fish that look
like snakes.

Other kinds of animals live in the ocean.

Starfish are ocean animals that look like stars.

This whale swims to the top of the water to breathe.

A sea otter plays near the **shore.**

Many plants and animals live in the ocean.

# Earth's Oceans

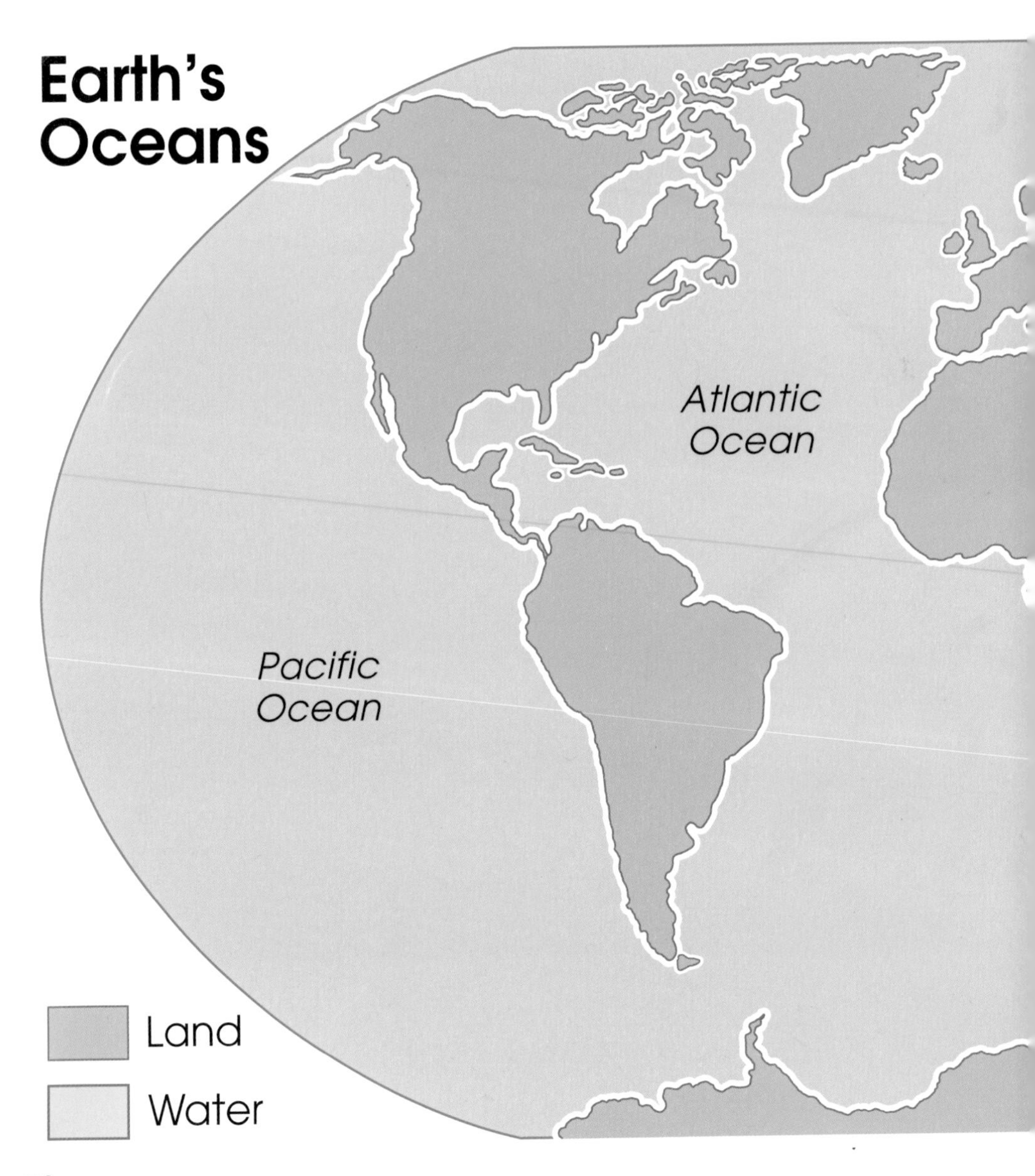

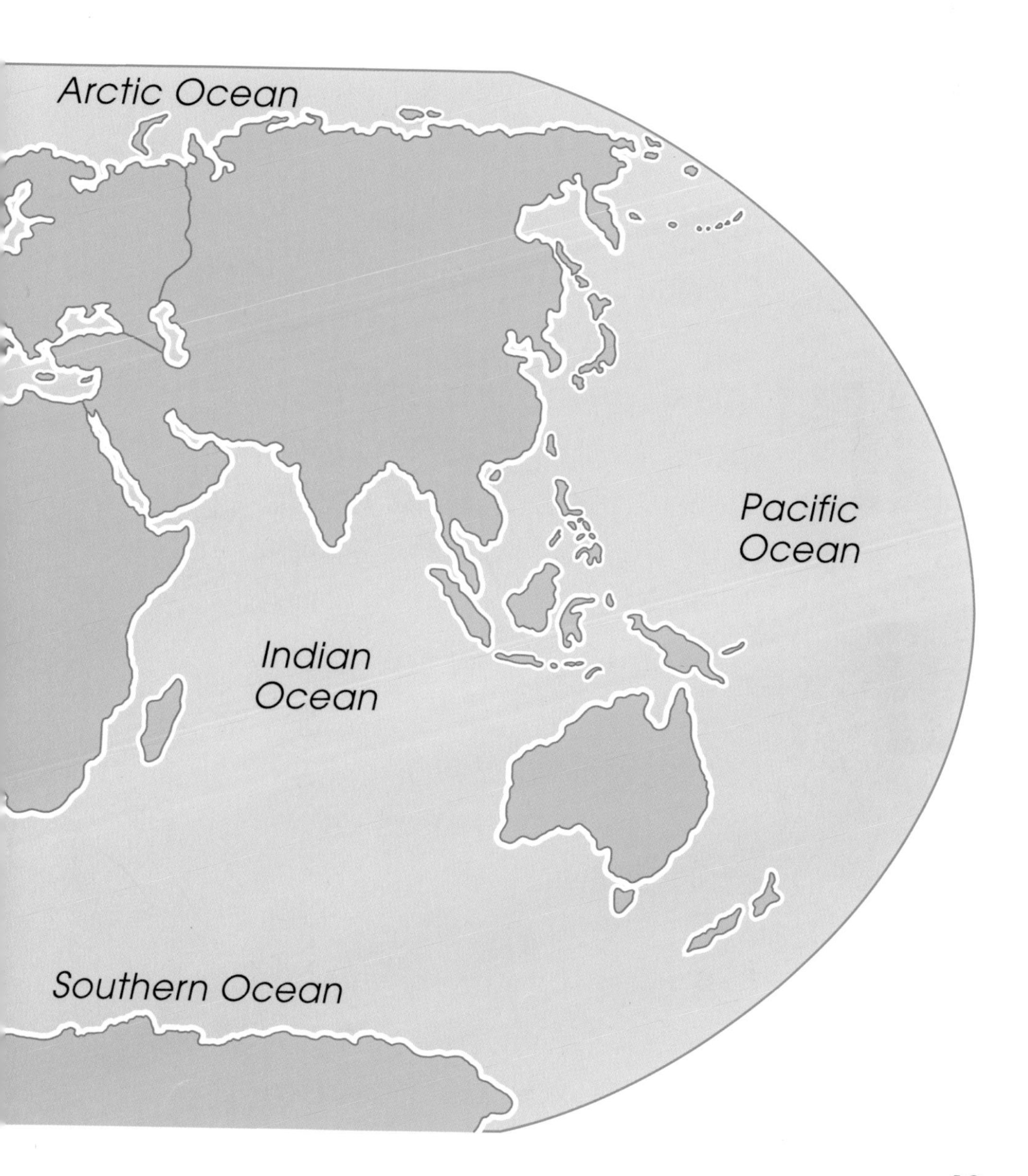
Arctic Ocean
Pacific Ocean
Indian Ocean
Southern Ocean

# Ocean Facts

Oceans cover nearly ¾ of Earth.

The five oceans are the Atlantic Ocean, the Pacific Ocean, the Indian Ocean, the Arctic Ocean, and the Southern Ocean.

Ocean water is salty.

Whales and dolphins breathe through blowholes on the tops of their heads.

The blue whale is the largest animal on Earth. It can grow as large as an airplane.

Sperm whales can hold their breath for more than one hour.

Green sea turtles can go up to five hours without breathing!

When a starfish loses an arm, a new arm grows in its place.

Sea otters use stones to open the shells of the clams they like to eat.

Penguins are birds that spend most of their lives in the ocean.

# Glossary

**fish** – a water animal with a backbone, fins, and gills for breathing underwater

**habitat** – where plants and animals live

**ocean** – a large area of salt water that covers nearly ¾ of Earth

**seaweed** – plants that live in the ocean

**shore** – land at the edge of the water

# Index

The photographs in this book are reproduced through the courtesy of: © Darrell Gulin/CORBIS, front cover; © Kent and Donna Dannen, pp. 2, 9, 22 (middle, bottom); © Guy Motil/CORBIS, p. 3; © Tom Bean/CORBIS, p. 4; © Novastock/Photo Network, p. 5; © Photodisc Royalty Free by Getty Images, pp. 6, 13, 14, 16, 22 (second from top);© Hal Beral/Photo Network, pp. 7, 8, 10, 11, 22 (top, second from bottom); © Kit Kittle/CORBIS, p. 12; © Royalty-Free/CORBIS, p. 15; © Howard Hall/Photo Network, p. 17.

Map on pages 18–19 by Laura Westlund.

Lerner Publications Company
A division of Lerner Publishing Group
241 First Avenue North
Minneapolis, MN 55401 U.S.A.

Website address: www.lernerbooks.com

Library of Congress Cataloging-in-Publication Data

Rivera, Sheila, 1970–
Ocean / by Sheila Rivera.
p. cm. — (First step nonfiction)
Includes index.
ISBN: 0–8225–2795–2 (lib. bdg. : alk. paper)
1. Marine animals—Juvenile literature. 2. Marine plants—Juvenile literature. 3. Ocean—Juvenile literature. I. Title. II. Series.
QL122.2.R58 2005
578.77—dc22 2004020789

Manufactured in the United States of America
1 2 3 4 5 6 – DP – 10 09 08 07 06 05